TABLE OF CONTENTS

19. Smart Fabric, Perfect Temperature: 1st Fabric that Automatically Adjusts to Environment
20. Smart Touch, Smart Phone, Smart Film: Alibaba's Humanitarian Smartphone Film
21. AI Smart Video Camera: China's Osbot Tail AI Camera
22. Self-Walking Suitcase that Follows You on its Own
23. Goodbye Toothbrush: France's Fasteesh
24. Latest Smartwatch with ECG: New Moves
25. China's Very Smartphone: Nubia's Red Magic Mars Smartphone
26. Exercising with Virtual Reality Gadget: Nordic Track's VR Bike
27. World's 1st Flexible, Bendable Smartphone: FlexPai Telephonic Innovation
28. Electric Skateboards, Green Travel: Boosted Mini
29. The Flying Suit: Gravity Jet Suit with 5 Mini Jet Engines
30. Hammock Needs No Trees: Freestanding Hammock-Tent Camping System
31. Flying Drone the Size of a Smartphone: Drone-X-Pro
32. New Wearable Heart Monitor: Mobile Heart Telemetry
33. Ford's New Smart Shopping Cart: Self-Braking
34. New Wrist Tech to Stop Allergic Reactions: EpiWear
35. New Radar – Tiny, Cheap, Effective: Tracks Up to 12 Miles
36. Wristband Changes Color if Drink Spiked: Innovation from Young Entrepreneur
37. Pure Water, Green Bottle: Purifies Water by Ultraviolet Light
38. AI Enhanced Hearing: Evoke's Machine Learning Technology
39. New, Full Body 3D Motion Tracking Suit
40. AI Beanie: Smart Hat to Helmet When Needed
41. Smart Curtin Removes Indoor Pollution: Air Purifying Textile
42. Scooter Sensation Stator: All Electric & Self-Balancing
43. Double Folding Smartphone: Phone Triples in Size

44. KitchenAid Smart Display
45. Smart Lighter to Quit Smoking: Slighter
46. Very Smart Clock, Smart Home: Lenovo Smart Alarm Clock Control System
47. Very Smart Mirror: Kohler's Verdera – Voice Activated, Motion Detection, Alexa

48. World's 1st AI Powered Pet Feeder
49. Communicative Cat Brush – Japanese Innovation
50. Touching VR Objects – New Ultra Light Gloves
51. Heart Monitor Taped to Skin: Organic Sensor & Solar Cell
52. LifeStraw: Personal Water Filtration Technology
53. Rockabye Baby in Robo Cradle: SNOO Smart Sleeper
54. High Tech Heated Clothing: Battery Powered Jackets to Stay Warm in Winter

55. 1st Rollable, Scrollable, Touchscreen Tablet: Inspired by Ancient Egyptian Scrolls
56. Robot Vacuum Cleaners Start to CleanUp

57. World's 1st VR Shoes and Gloves: Japan's Wearable VR Innovation
58. Paper Water Test – Quick, Cheap, Green: Biosensor ID's Water Quality

59. World's 1st Biodegradable Cooler: No Styrofoam, Environmentally Friendly

INTRODUCTION

This book is loaded with the latest new gadgets just invented around the world. These are exciting new innovations that answer long-standing practical needs. They are gadgets you'll love using and will trigger the thought "Why didn't I think of that?"

We showcase smart mirrors that superimpose new clothing outfits on your image, so you don't have to try them on. Tiny, robotic teeth cleaners that whiten your teeth and remove plaque. A gadget called CLIP that turns any bike into an electric bike in seconds. A breakthrough, washable, wearable display for your electronics that's solar powered. Samsung's Smart Shirt that's wearable technology to monitor your heart. Nordic Tracks' new Virtual Reality bike that enables you to exercise while wearing a VR gadget. A vertical, indoor hydroponics garden iHarvest that grows up to 30 plants year-round. We spotlight a number of new travel and sports gadgets, including:

- The flying suit with 5 mini-jet engines to lift you off
- Electric skateboards that travel up to 20 mph
- New, full body, 3D motion tracking suits for athletes and others
- Self-walking suitcases that follow you
- Scooter sensation Stator, that's all electric and self-balancing
- AI Beanie that goes from soft to helmet hard when needed.

Other gadgets that you would probably love to have include the

world's 1st AI powered pet feeder, smart lighters to quit smoking, smart clocks that serve as control systems for your smart home, smart curtains that remove indoor pollution and Kitchen Aid's smart display. The list goes on.

"List of the Latest, Top Gadgets" is a fun news read about some of the most innovative and fun gadgets just invented.

AUTHOR'S BIOGRAPHY

Ed Kane is the author of 9 books on innovation including "List of Best New Innovations" and "Important Innovations Collection". He created and serves as Executive Producer of CEO Global Foresight, a national program on PBS focused on innovation. Ed is a science graduate of the University of Pennsylvania. He is an avid researcher into the future of breakthrough innovation and its impact on humanity.

1. Smart Mirrors Superimpose Outfits on Your Image

Source: MemoMi

Artificial Intelligence Changing Retail: No More Fitting Rooms

The retail industry will be dramatically changing over the next ten years. Artificial intelligence, augmented reality and virtual reality are expected to radically transform the shopping experience. In fact, it's already started.

Smart Mirrors

Smart mirrors are being deployed at nearly three dozen Neiman Marcus stores. Using AI, AR and gesture recognition technology, the digital mirrors superimpose clothing on your image. It's a virtual changing room that lets you see yourself in different outfits in a matter of seconds. No need to change in a fitting room.

MemoMi

The smart, digital mirror was invented by the company MemoMi, whose founder and CEO Salvador Nissi Vilcovsky holds 20 patents. The smart mirror can be used for make-up, eyewear, footwear, clothing and accessories in real-time. You can even share your image on social media to get reaction from your friends. This is just the beginning of new technologies transforming the shopping experience.

2. Very Smart Pill Bottle from Saudi Arabia

Source: KAUST

Sends Wireless Alerts if Tampered with or Unsafe

A team of engineers at KAUST - King Abdullah University of Science and Technology - have developed a smart pill bottle that is loaded with highly advanced technology. The bottle sends a wireless alert on tampering, overdose or unsafe storage. The alert is sent by a stretchy mobile sensor, which is a flexible computer.

Low Cost Sensors Protecting Patient Medication

The sensor is low cost and composed of two layers of copper tape with silver particles between them. When the material is pressed, the copper layers make contact and send an electrical signal alert to an external reader.

Much More Tech in this Medical Gadget

The 3D printed lid counts the pills using light-emitting diodes. An adhesive sensor monitors humidity and temperature inside the bottle. Finally, the bottle is sealed with touch sensitive tape. The sensor array detects if someone has opened the bottle and if the interior of the bottle has become dangerously moist. Either triggers an alert sent by Bluetooth to cell phones.

3. Navigation Display on Car Dashboard: VIZR & Smartphone Advanced Display

Source: VIZR

Heads-Up Display Gadget

VIZR is an advanced navigation display that grips to the car dashboard. You put your smartphone on VIZR and it turns your smartphone into a transparent display that lets you navigate and drive while never taking your eyes off the road. The heads-up display (HUD) uses the same technology that fighter pilots use to stay focused. The system offers a number of apps to view such as navigational tools like GPS, street maps, traffic conditions and speed.

Distracted Driving

Over 424,000 drivers are injured yearly by distracted driving. For instance, by looking back and forth between a screen on the car seat and the road. This new piece of smart gadget technology is designed to change the dynamic and enable the driver to navigate and keep their eyes on the road for greater driving safety.

4. Lost & Found Tracking Gadget: XY Find It

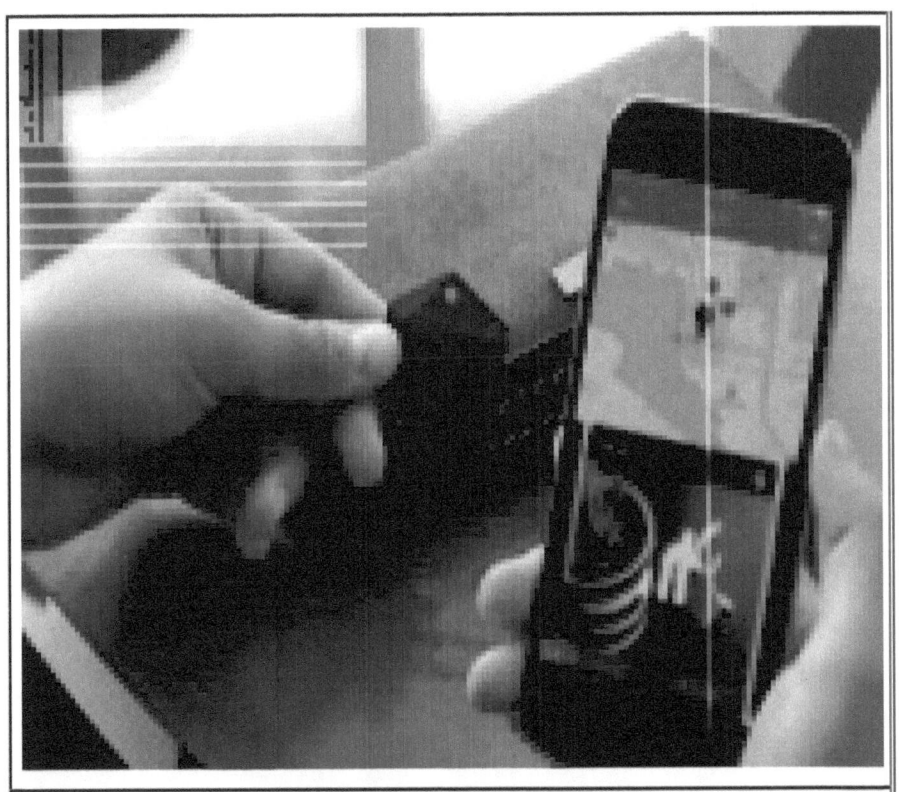

Source: XY Find It

Global Tracking Network

XY Find It technology enables you to find a lost item up to 300 feet away within seconds. The company adds, with its global network of systems deployed, you'll find anything you've misplaced in the world very quickly. The apps pick up signals from lost, tagged items and notify the owners. It uses Crowd GPS technology. XY4 is a coin tag that you attach to anything you don't want to loose. Examples include your keys, dog, wallet, car, purse - you name it. The coin tag along with the XY Find It app will locate the object in seconds. You can log into the app for the exact location.

A Lot of Technology in this Gadget

If you can't use the app because you've lost your phone you can press XY4* to make the phone ring very loud, even if it's on silent, to locate it. The system has technology to notify you that you're walking away from an object you don't want to lose. The battery lasts for up to 5 years. And the company says it's a lot less expensive than GPS.

5. Car Monitor Prevents Unnecessary Repairs: FIXD - Car Health Monitor

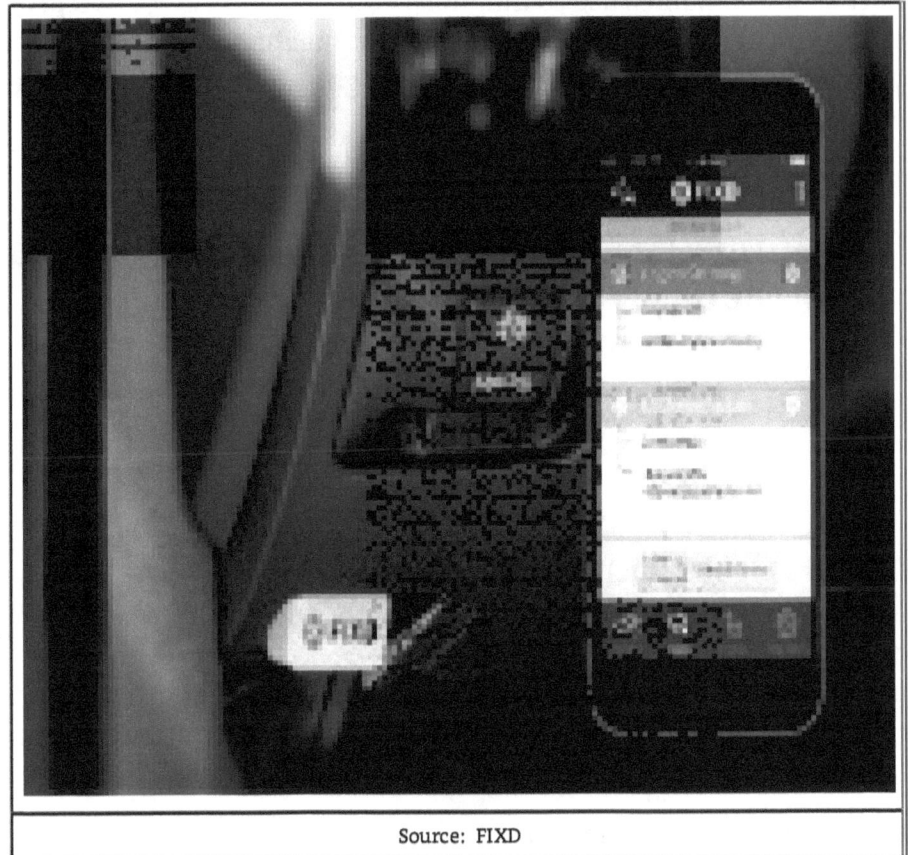

Source: FIXD

New Technology That Puts You in the Driver's Seat

This monitoring device was invented by three engineering gradu-
ates from Georgia Tech. They say it's a smart gadget that can save
you thousands of dollars in needless car repairs. FIXD is the first,
easy to use car health maintenance monitor. When a car warning
light goes on, the device tells you if there is a problem, how severe
it is, if it's an emergency and how much it will likely cost to repair
the problem.

Information is Power

FIXD fits in below your car's steering wheel in the onboard diag-
nostic port and sends the analysis to your phone via an app. Every
car built since 1996 is required to have that port. It's the same

port that mechanics use to diagnose your car. But now, FIXD provides you with the readout to tell the mechanic exactly what your car needs or doesn't need. You might say it puts you in the driver's seat.

6. Ultrafast Camera Images Speed of Light: 100 Billion Frames Per Second

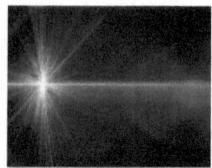

Captures Light Travelling from Point to Point
A researcher at Washington University in St. Louis has invented a camera that captures 2 dimensional images at 100 billion frames per second. That speed is capable of shooting phenomena that previously were too fast to be observed such as light pulses and laser beams as they travel from one point to another.

World's Fastest 2 D Camera
The new camera was invented by Professor of Biomedical Engineering Lihong Wang of Washington University. His research was supported by the National Institute of Biomedical Imaging and Bioengineering. Wang's camera is the world's fastest 2 D camera that doesn't require an external flash or multiple exposures. It's a great way for imaging ultrafast phenomena like a laser flash or biochemical reactions.

Teaming the Camera with a Microscope for More Scientific Discoveries
Wang is now working to pair the camera with a microscope to capture previously unobservable biological events. An example is how laser light might destroy diseased tissue but leave healthy tissue unharmed.

7. E-Friendly Paddleboard: New Modular Board that's Good for the Environment

Source: Easy Eddy

Easy Eddy

Easy Eddy is a modular, standup paddleboard that breaks into three parts for easy handling, transport and storage. It's specifically designed to overcome the cumbersomeness of bulky traditional boards. . Massachusetts based startup Artech's water paddleboard is a new, green innovation. It's made from reclaimed and recyclable polyethylene, making it good for the environment. The board is ten feet long and takes less than 60 seconds to snap together.

Enjoying the Environment and Not Hurting the Environment
Easy Eddy weighs 37 pounds and fits into most cars, SUVs and boats. And it can support weights up to 200 pounds. Among the features included with the board are an action-cam mount, full length deck pad, sealed storage compartment, water bottle receptacle and bungie straps to secure items onboard. Artech is in the kickstarter phase with Easy Eddy. The planned retail price is $1499. The invention is the result of new technology creating a green, clean paddleboard.

8. New Water Sport Weeebo: Powered Board "Water Gadget"

Source: Yanmar's Wheeebo

From Japan at Speeds Up To 3&1/2 MPH

This looks like a fun new piece of technology. It's a powered standup board to skim across the water. It's called the Wheeebo and was invented by Japanese manufacturer Yanmar, which is well known for marine engines. Yanmar calls it an "on-water gadget".

Stand-Up Board

The Wheeebo is a round, saucer like board that is 145cm in diameter. It has a battery powered engine. The user steers it by shifting their weight on the board to go in the direction they want. Sensors detect the weight shift and adjust propellers under the board. Top speed is 3&1/2 mph. There are two acceleration speeds which the user controls by a hand held remote.

Calm Waters

The Wheeebo is designed for lakes and bodies of water that are relatively calm. The company says it's very easy to use. Yanmar intends to continue testing it this year and make it available to consumers in 2020. The battery on the board lasts for an hour before needing a recharge.

9. Smart Contact Lens: DARPA May Have Found It

Source: DARPA

Giving Soldiers "Superpowers"

The technology comes from France's engineering school IMT Atlantique. They've created the first standalone contact lens with a tiny, flexible microbattery. The US Defense Departments Advanced Research Projects Agency DARPA is interested in the technology to enhance soldiers' visual capabilities in the field and a lot more. This appears to be the answer to DARPA's 10 year search for very smart contact lens.

Breakthrough Smart Visual Technology

This is a wirelessly connected contact lens capable of providing augmented vision assistance and relaying visual information wirelessly. IMT has developed a flexible battery that continuously powers a LED light for several hours. They add that graphene-based flexible electronics can enhance the smart contact lens' capabilities.

Many Applications

Besides greatly augmenting soldiers' capabilities in the field, this has applications for surgeons in the operating room and drivers on the road. Besides DARPA, Microsoft is said to be interested in making a major investment in the technology. IMT hopes to start testing it and going into clinical trials in 2020.

10. Squishy Robots from UC Berkeley: Disaster Bots

Source: Squishy Robotics

For Search and Rescue Missions

Round, squishy robots capable of being dropped 600 feet and surviving have been developed by engineers at UC Berkeley and Squishy Robotics, a startup company founded by UC Berkeley Engineering Professor Dr. Alice Agogino. The robots are designed for use in disaster areas to collect information on the ground.

Space Bots

These "tensegrity" robots were originally designed to explore Saturn's moon Titan. They would be dropped from a spacecraft onto Titan. The UC Berkeley team realized they could be used in disaster areas on earth by equipping them with sensors able to detect conditions like dangerous gases. The robots relay that information to first responders.

Rapid Deployment

The sensor robots can be rapidly deployed, are mobile and, according to Dr. Agogino can work as co-robots with their human partners at disaster scenes.

11. Tiny Robotic Teeth Cleaners: Army of Microrobots Wipe Out Dental Plaque

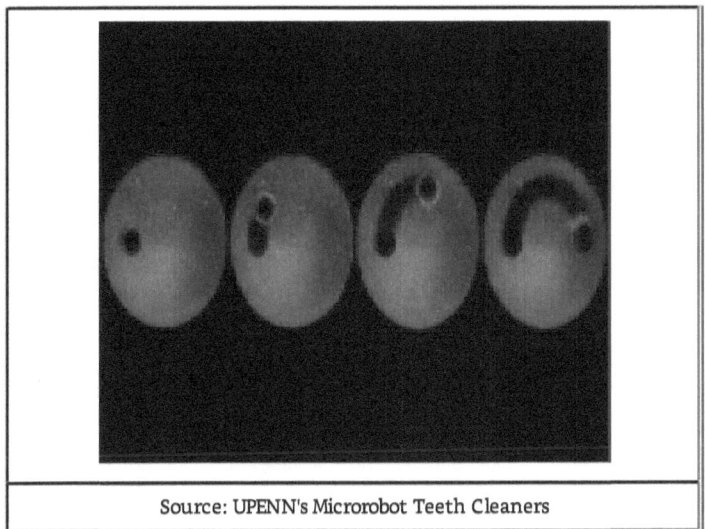

Source: UPENN's Microrobot Teeth Cleaners

New Innovation from the University of Pennsylvania

A team of engineers, dentists and biologists at the University of Pennsylvania have developed a crew of microscopic, robotic cleaners for your teeth. The new technology combines two robotic systems: one for the surface of the teeth and the other for inside, tough to get to places between the teeth and inside the teeth such as for root canals. Both robotic systems rely on iron-oxide nanoparticles.

CARs

The UPENN researchers call their system CARs, or Catalytic Antimicrobial Robots. The microbots can be steered magnetically to breakdown biofilm and accumulated material like plaque on teeth. This is a robotic biofilm removal system that has a wide range of applications. The team continues to work on their technology. They believe other applications include keeping water pipes and catheters clean as well as reducing tooth decay, endodontic infections and teeth implant contamination. This important new innovation has been published in the journal Science Robotics.

12. Vertical Indoor Gardening: iHarvest Grows More with Less Indoors, Year-round

Source: IGWorks

How Does Your Garden Grow?

This is patent-pending hydroponics technology with a lot of green promise. It's called the iHarvest vertical garden. It's an indoor, hydroponic garden that is fully automated. The inventor is IG Works (Indoor Garden Works). The company is a start-up. The hydroponic system includes a trellis support and full spectrum LED lights. It's designed to look attractive in the home while providing fresh fruit and vegetables year-round.

Green, Clean and Easy

The system can grow 30 fruits and vegetable plants simultaneously. The company says it needs only 2&1/2 feet of floor space. They add it grows the plants three times faster and 30% larger than traditional outdoor gardens do.

Indoor Garden at Your Disposal

iHarvest uses no pesticides. The company claims it uses 95% less water and 60% less fertilizers than outdoor gardens require. It is fully automated, easy to maintain, fits into the smallest apartment and most importantly promises to give you fresh, pesticide-free fruits and vegetables year-round.

13. Self-Balancing Electric Bike: Very Smart Motor

Source: Delft University of Technology

Innovation from the Netherlands' TU Delft

This is the first system in the world that can help a bicycle stay upright. It was designed by researchers at Delft University of Technology and bike manufacturer Koninklijke Gazelle in the Netherlands. They've designed a smart steering system that may reduce the number of accidents when cyclists loose balance and control of their bikes.

Smart Motor, Smart Sensor

The bike has a sensor that tracks the bike's movements and determines if it's in danger of tipping. If the system determines that a fall is about to happen, the electric motor in the steering tube activates and turns the handlebars to balance the bike and prevent a fall.

Tipping Point

The researchers say their bike smart steering system is a prototype. They want to do more research on it before making it available to consumers. Specifically, they want to maximize the comfort level for cyclists when using the system. The researchers had seniors in mind when designing the system to help them stay active. But it works for anyone. Electric bikes are involved in more falls as they are faster and heavier than regular bikes.

14. CLIP Turns Any Bike into Electric Bike: Portable, 4 Pound Clip-on Electric Motor

Source: CLIP photo

CLIP on to Cycle Electric

The portable electric motor CLIP turns any bike into an electric bike in seconds. The device is small enough to fit into a backpack and weighs just four pounds. It attaches to the front fork of the bike easily in seconds.

Rechargeable at Any Power Outlet

The device's battery powered, 450 watt motor is designed to add a boost of power to get up, for instance, difficult hills. The inventors hope it makes bike commuting much more doable to cut CO_2 emissions from cars. It's also bike share compatible. Co-founder of CLIP Somnath Ray has an MS from MIT and worked at the MIT Media Lab on urban mobility. He commuted by bike to work at the Lab and envisioned his bike going electric.

15. Smart Wrist Band: Stroke Early Warning System

Source: Kaunas University of Technology

International Biomedical Engineering Breakthrough from Europe

It looks like a watch, but it's an early warning system that detects the early stages of atrial fibrillation and stroke. It could help prevent stroke and other health complications. It has just been developed by an international team of biomedical engineers and scientists from Lithuania's Kaunas Institute of Technology and Sweden's Lund University. It's a very smart wrist band that could save lives.

Prototype

This device is still in the prototype stage. But it has great potential. The smart wrist band features both PPG and ECG sensors. The PPG sensor optically senses changes in blood volume in the tissue. That can be a sign of the start of atrial fibrillation. If a blood volume change is detected, that sounds an alert to the wearer to touch the device with their other hand to register an ECG signal reading. The ECG sensor measures the heart's electrical activity via the skin.

Algorithm Reading

The two readings are analyzed by algorithms to determine if a-fib is about to begin. This provides potentially life-saving time for the wearer to contact a doctor and get help sooner rather than much later. The wrist band monitors blood flow continuously for any signs of a-fib. Until now, people at risk have been monitored by periodic trips to a clinic.

16. New Washable, Wearable Display: Solar Powered Wearable

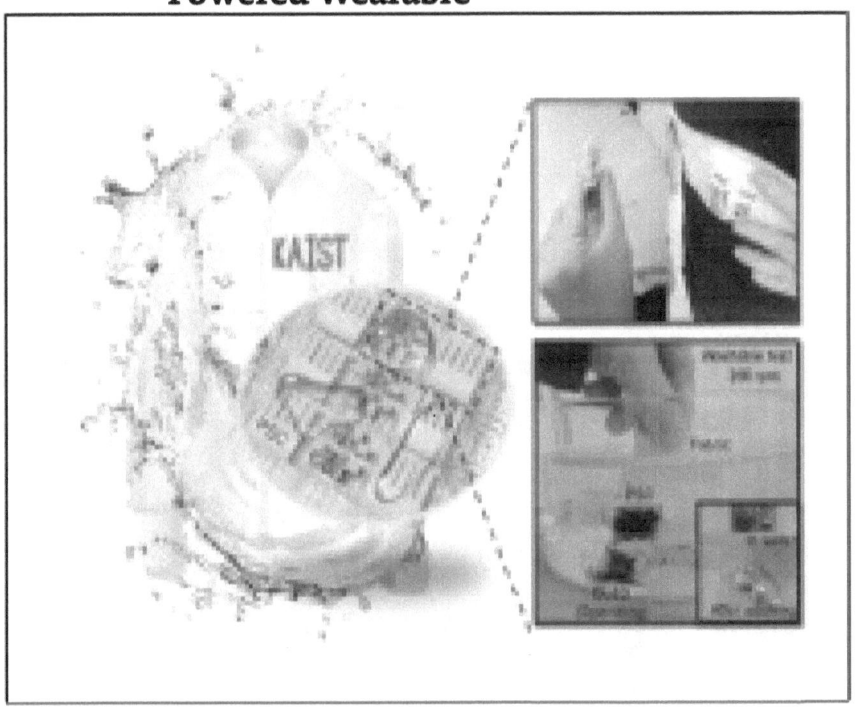

Source: KAIST New Wearable Display

From South Korea's KAIST

A South Korean electrical engineering team has developed a machine washable, wearable textile display that is self-powered by solar energy. The wearable contains polymer solar cells (PSC) with organic light emitting diodes. The inventors are electrical engineers at the Korea Advanced Institute of Science and Technology (KAIST). This is a breakthrough innovation in wearables for electronics.

PSC

The polymer solar cells provide stable power without any external power source for the wearable display. KAIST engineers consider PCS the most promising next generation power source especially for wearable electronics.

Machine Washable

To protect the electronic textile from moisture and oxygen that can damage them, the KAIST team improved on the material barrier using spin coating and atomic layer deposition. The result: a wearable textile that's machine washable with little change in performance.

17. Microneedle Patch Kills Infection, Delivers Vaccine: Safer for Developing World

Source: University of South Australia microneedle patch

Silver Nanoparticles Kill Bacteria

Scientists at the University of South Australia have developed a microneedle patch embedded with silver nanoparticles and vaccine. It offers an alternative to traditional needle injections and eliminates infections at the injection site.

Alternative to Traditional Needles

The patch carries an array of microneedles that contain both vaccine and silver nanoparticles. When attached to the skin, the microneedles dissolve in one minute. The patch is painless, doesn't need to be refrigerated and releases the vaccine into the top layer of skin, not reaching any nerves. The vaccine is released into the bloodstream and the silver nanoparticles kill any bacteria. Bottom-line: it enables safer vaccinations and prevents infections. The team believes it would be particularly effective in the developing world where vaccinations with unsterile needles result in infections.

18. Samsung's Smart Shirt: Wearable Tech to Monitor Your Health

Source: Samsung's Smart Shirt

Shirt Sensors Transmit Results to Your Smart Phone

This is cutting edge, patented, personal health care technology. Korean manufacturing giant Samsung has invented a smart shirt with embedded sensors that detect early signs of respiratory problems and lung disease. The resulting diagnosis from the sensors is transmitted to your smart phone for you to see. The company says the possibilities of health monitoring by this type of wearable technology is unlimited. Consider the benefits to individuals, the elderly, children and athletes to know there's a problem and seek immediate medical attention.

Samsung About to Launch Smart Shirts

The smart shirt's sensors listen to the wearer's breathing to detect signs of respiratory problems. The wearer's health history, age, height, weight, gender and other variables are taken into account. The smart shirt with health diagnosis sensors can spot colds, pneumonia, asthma, bronchitis, respiratory problems and lung disease. The system also sends a recommendation to your smart phone, such as "see your doctor". Samsung is readying to launch but no target date or price has been provided.

19. Smart Fabric, Perfect Temperature: 1st Fabric that Automatically Adjusts to Environment

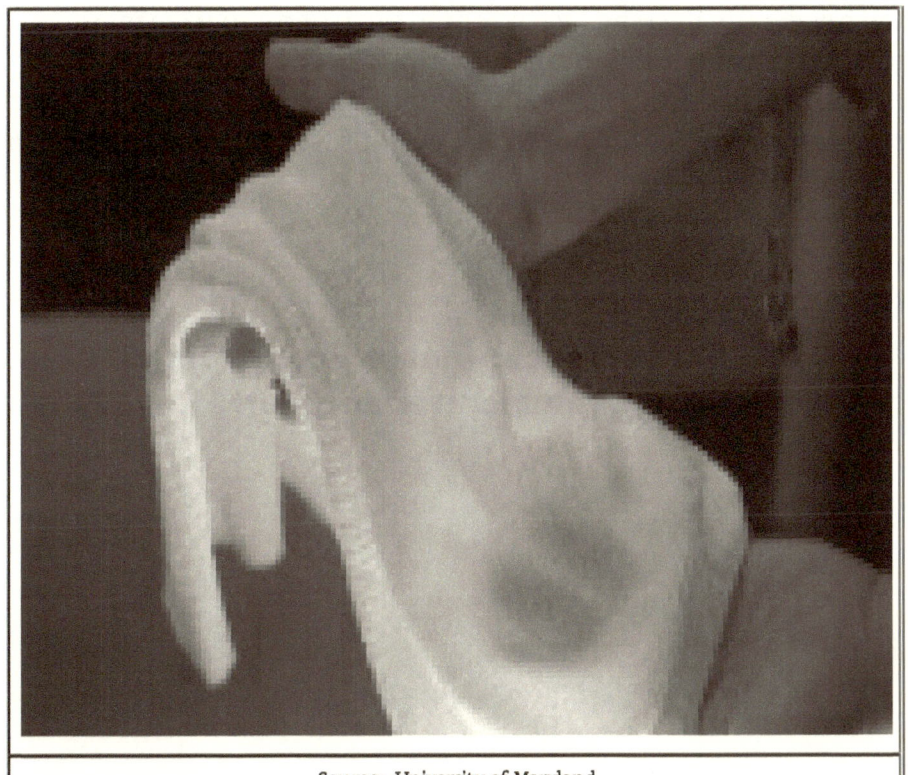

Source: University of Maryland

New Temperature Adjusting Fabric

This is a world first. A wearable material that automatically adapts its insulating capabilities to changes in its environment. This is pioneering research toward the commercialization of comfort adjusting clothing.

Gates Heat and Cold

University of Maryland researchers have created the new fabric. The material "gates" heat and cold. According to the research team, this is the first technology to dynamically gate infrared radiation. To put it simply, the fabric let's heat go out if it get hot and holds it in if it's cold. The scientists say the process is so fast, you're not aware of changing temperature conditions.

Smart Fabric

To automatically adjust temperatures, the fabric is engineered with specially designed yarn coated with a conductive metal. To take this commercial, more research is needed and the manufacturing process needs to be scaled up. But what the University of Maryland team has invented is a world first. It was published in the journal Science.

20. Smart Touch, Smart Phone, Smart Film: Alibaba's Humanitarian Smart Phone Film

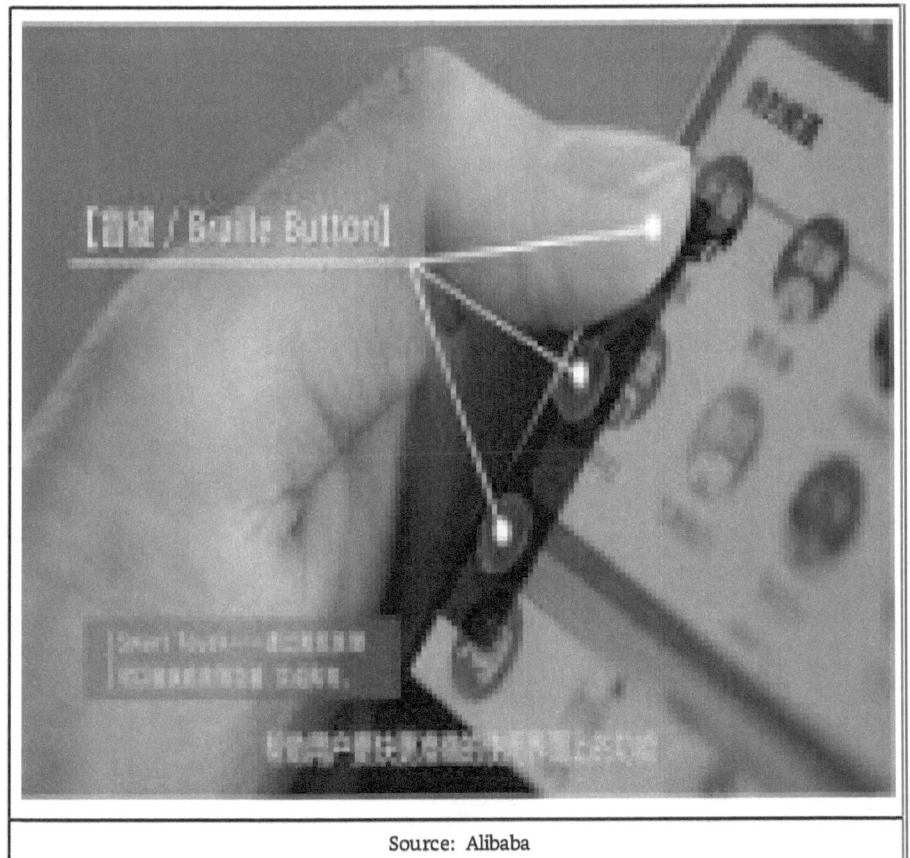

Source: Alibaba

Enables Visually Impaired to Stay in Touch, Shop, Pay Bills

This is great, humanitarian innovation. China's technology giant Alibaba has invented a tactile, thin film interface for smartphones to work for visually impaired users. It's called Smart Touch and adds a tactile button to smart phones. The Smart Touch film attaches to the smart phone display.

Three Small Buttons

The film has three buttons including braille, confirm call and go back. These are short cuts for the visually impaired to better use the phone. Alibaba designed it to be affordable and manufacturable at less than 4 cents to produce.

Retail Distribution in Early 2020 With No Profit

Alibaba doesn't intend to make a profit on this new technology. It's going to launch the smart film in retail stores in 2020. It is working and testing it right now and wants to make its smart phone device agnostic, meaning you can use it on any smart phone. The company believes it will make shopping and doing payments on line a lot easier for the visually impaired.

21. AI Smart Video Camera: China's Osbot Tail AI Camera

Source: Obsbot Tail AI Video Camera

Highly Advanced Videographer
The Obsbot Tail AI is a brand new video camera that packs a lot of cutting edge technologies including artificial intelligence.

Automatic, Smooth Video Tracking
The camera functions like a professional is operating it. Specifically, the AI identifies subjects of the shoot and then automatically

tracks them with smooth camera movements. The AI recognizes the face, body shape and posture of the person, resulting in highly accurate tracking, camera pans, tilts and zooms. It shoots up to 150 minutes on a charge. The videos are up to 4K/60fps and also shoots still, 12 megapixel photos.

Gesture Control

The camera can be operated by gestures, which allows the person in front of the camera to go back and forth between different modes. It contains advanced imaging technology. And essentially provides anyone who wants to produce videos the technology to do so without guidance from someone behind the camera. It's getting great reviews.

22. Self-Walking Suitcase that Follows You On its Own

Artificial Intelligence Luggage

This suitcase looks like an ordinary piece of carry-on luggage. But it has a name - Ovis - and is loaded with artificial intelligence that enables it to follow you around, wherever you go, on its own.

Making for a Perfect Trip

This is a very smart suitcase. It has artificial intelligence enabled sensors all over the exterior. That enables it to follow its owner around on its own.

ForwardX Robotics of China

The Chinese AI startup company ForwardX Robotics innovated this unique travelling machine. The company says it will retail for $799. and can be pre-ordered in 2019.

23. Goodbye Toothbrush: France's Fasteesh

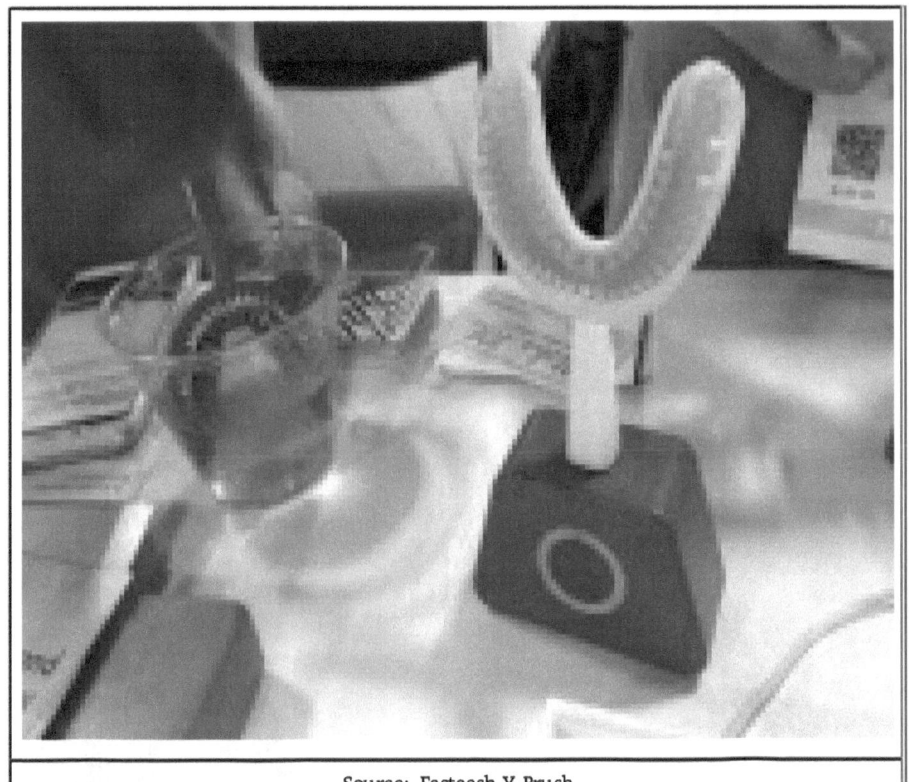

Source: Fasteesh Y-Brush

The Innovative Y-Brush
Imagine cleaning your teeth in ten seconds instead of the several minutes it takes with an ordinary toothbrush. Fasteesh, a French company and creator of the Y-Brush, claims you can achieve the same level of clean with their new product.

Mouthguard
The Y-Brush is a mouthguard that can fit around your upper and lower rows of teeth. The Y-Brush is all nylon bristles placed at 45 degree angles in the mouthguard. It takes just 5 seconds to clean each of the two rows of teeth.

Smaller Sizes for Kids
Awaiting approval by the American Dental Association, it's expected to ship during 2019. The likely prices is $125.00 The Y-

Brush uses ordinary toothpaste and comes in four sizes, including smaller sizes for children.

24. Latest Smartwatch with ECG: New Moves

Source: Withings Move ECG

Withings Move ECG

This watch has been named by PC Magazine the Best Wearable out of the Consumer Electronics Show 2019 in Las Vegas. It's called the Withings Move ECG and was developed by the United Kingdom based Withings company.

Built in Electrocardiogram

This is the first analog smartwatch with an electrocardiogram built into it. PC Magazine says, it gives the Apple Watch Series 4 a run for its money.

Simple and Inexpensive

The ECG measurement is simple and the price is $129. The watch runs for a year on a single battery. It looks like a great way to monitor your health while timing your run.

25. China's Very Smartphone: Nubia's Red Magic Mars Smartphone

Source: Nubia Red Magic Mars

Coming to US 2019

The Nubia Red Magic Mars is being called great tech and great value. It was showcased at the Consumer Electronics Show 2019 and PC Magazine made it their top phone pick from CES. China based Nubia has committed to selling it in the US and North America for $399 starting in 2019. Nubia is a subsidiary of the Chinese phone company ZTE.

Smartphone and Gamer's Phone

The Nubia Red is equipped with a Snapdragon 845 chipset, a 6" 1080 p display and the highest end model has 10 gigs of RAM. The smartphone is also geared to gamers. The phone has a 16-mega-pixel rear camera, a fingerprint scanner and a strip of LED lights that display in different colors. PC magazine reports its perform-ance is as good as that of $700 competitors.

26. Exercising with Virtual Reality Gadget: NordicTrack's VR Bike

Source: NordicTrack

VR Cycling
This exercise bike was unveiled by NordicTrack at the Consumer Electronics Show 2019 in Las Vegas. PC Magazine has named it the Best Health and Fitness Device released at the CES.

VR Headset
The HTC Vive Focus headset lets you enjoy VR gameplay while you're burning calories peddling on the bike. And the exercise and VR are in sync. Uniquely, the incline on the bike automatically adjusts to the elevation of what you're seeing on the headset. The process is automatic and in sync.

VR Exercise
This is a brand new, mind over matter approach to exercise. In fact, it can help you forget that you're exercising. It is pricey: $2,000 which includes the bike, headset and a 1 year membership in iFit for exercise and gaming content.

27. World's 1st Flexible, Bendable Smartphone: FlexPai Telephonic Innovation

Source: Royole FlexPai

Tablet and Phone In One

The world's first, flexible, bendable, foldable smartphone debuted at the 2019 CES tech show in Las Vegas. The phone folds out into a tablet and back into a phone. It can fold from 0 to 180 degrees. It's been under development for five years by a company named Royole based in Fremont, CA and is called the Royole Flex-Pai.

Cool Tech

This is a developer's model and demonstrates the future of smartphones. The operating system that runs on top of the Android system is named Water. It automatically changes the screen as you flip the phone. The company says the screen has been resistance tested and can be folded tens of thousands of times without any problem. It's getting good reviews for being cool tech but a few reviewers have said it's a little heavy and bulkier than a traditional cellphone. The company says it can fit into your pocket. It's a world first under development. The developer model, pictured above, can be ordered for $1318.

28. Electric Skateboard, Green Travel: Boosted Mini

Source: Boosted Mini

"A Tesla in Your Backpack"
It's a 20 mph alternative to travelling a short distance in your car. The electric skateboard boosted Mini weighs 15 pounds, is 30 inches long and can maintain speeds of 20 mph. Skateboards are known for recreation and fun. The new electric models are being marketed as a means of transportation. As the CEO of Boosted Jeff Russakow puts it "It's a Tesla in your backpack." and a step toward

a less car dependent future.

Not Inexpensive

Boosted Mini costs $750. One issue to be aware of with regards to electric skateboards is the high capacity batteries that power them have had overheating problems in the past. But Boosted Mini has been cited by Time Magazine as one of the top 50 inventions of 2018. It's light and small enough to carry into your destination.

29. The Flying Suit: Gravity Jet Suit with 5 Mini Jet Engines

Source: Gravity Industries' Jet Suit

Dressing for Altitude

The Gravity Jet Suit has been cited as one of the best innovation inventions of 2018. London based Gravity Industries invented and is now developing it.

Plenty of Lift

The 1,050 horsepower system is powered by 5, mini jet engines. One engine is built into a backpack. The other four jet engines are attached to the hands. The system can achieve speeds of 50 mph. The ability to fly it takes a lot of learning, practice and experience.

Expensive Flights

A recent Gravity Jet Suit cost $440,000. As they are powered by jet engines, the suits are very loud. But innovator/inventor Richard Browning hopes to raise money from public events like races to drop the price and then convert the Gravity Jet Suit into a cheaper, electric flying suit.

30. Hammock Needs No Trees: Freestanding Hammock-Tent Camping System

Source: LIT Outdoors

Tammock - New Camping Invention
This new gadget was just introduced by the company LIT Outdoors. They call it the Tammock. It's a free standing invention: a combination Hammock-Tent system that brings the comfort of swinging in a hammock to outdoor areas without any trees. Or as the company puts it "brings it all down to earth".

Hammock Camping System
There are other tent-hammocks on the market but they require trees to be setup. The Tammock has a freestanding tent and hammock that can be setup anyplace. It has a waterproof rainfly providing tent protection from the rain. The doors on each side of

the tent are rollups. It's an all in one hammock camping system, good for the beach, desert, the mountains and your backyard. It's a brand new invention that is in the Kickstarter phase.

31. Flying Drone The Size of Smartphone: Drone-X-Pro

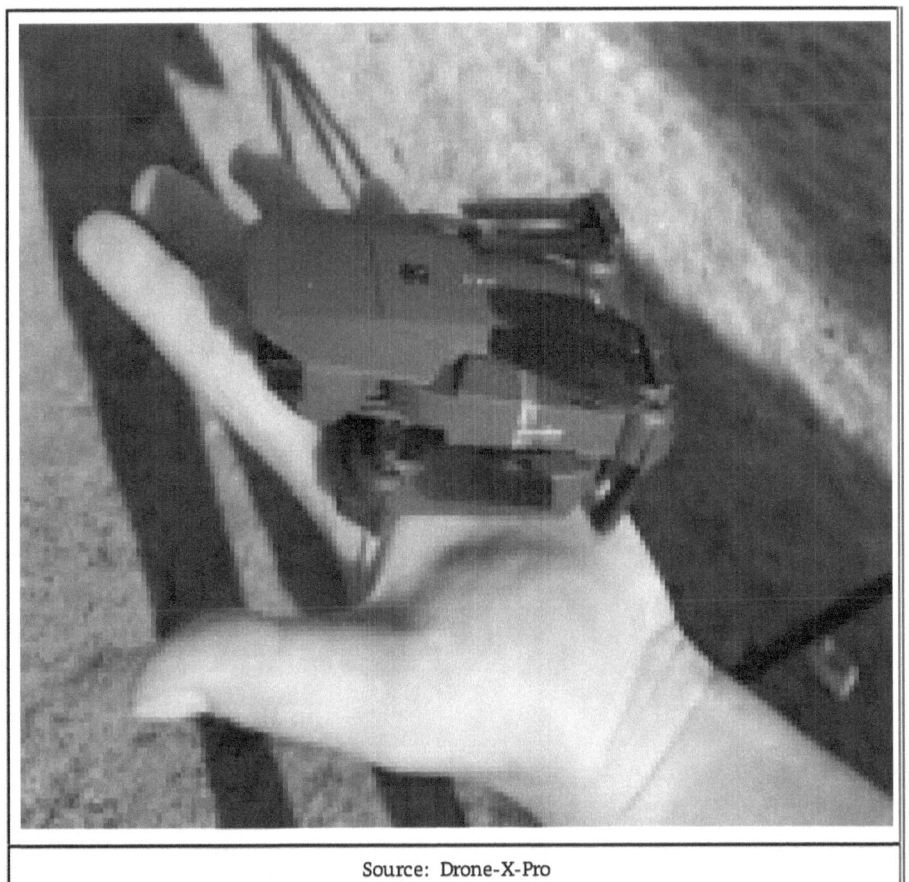

Source: Drone-X-Pro

Easy Flying - Tiny Drone - Smart Gadget

If you're interested in drones, the inventors of the Drone-X-Pro say this small drone is easy enough for anyone to operate. It's the size of a cellphone that you can hold in the palm of your hand.

You charge the battery, install the App and in under ten seconds, you're good to go. You control the small drone's flight by your smartphone.

Foldable Propellers
This is a precision engineered drone specifically designed for easy flying inside and outside. Its propellers fold in for easy carrying and the design is lightweight.

HD 360 degree Images
The drone is equipped to take HD photos and videos and can capture 360 degree images from the air. It's equipped with gravity sensors to automatically avoid collisions with objects and the ground. Drone-X-Pro can fly up to 12 meters per second and a maximum distance of 2 km. And, it's capable of flying ten minutes without having to land and change batteries. It's a fascinating example of a tiny drone that's a smart flying gadget.

32. New Wearable Heart Monitor: Mobile Heart Telemetry

Rhythm Express RX-1
The Rhythm Express RX-1 is an advanced, remote cardiac monitoring system invented by VivaQuant of Minneapolis. It has just been cleared for use by the FDA. The wearable, wireless heart monitor provides heart patients continuous coverage and gives them more freedom. It monitors a patient's arrhythmia with its artificial intelligence and patented wavelet-based analytics, 24/7 at home, work, wherever they are.

Mobile Cardiac Telemetry

The one piece device also functions as a Mobile Cardiac Telemetry unit, wirelessly transmitting the readings for remote analysis by medical staff, allowing the patient to be monitored and diagnosed no matter where they are. The inventor, Dr. Marina Brockway says it provides faster, higher quality diagnosis at a lower cost. It works continuously for two weeks without a recharge.

33. Ford's New Smart Shopping Cart: Self-Braking

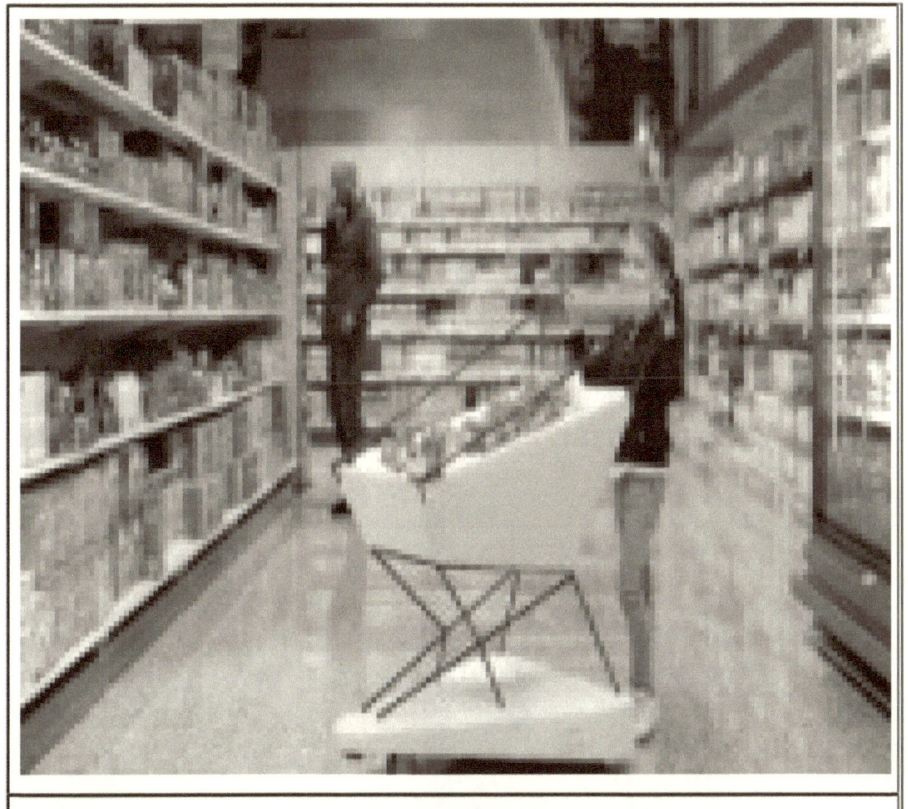

Source: Ford Europe

No More Runaway Carts at the Grocery Store
Runaway shopping carts pose a danger to cars and customers at grocery stores. Ford has invented a self-braking shopping cart that stops automatically to prevent damage.

Pre-Collision Assist System
The cart has a version of Ford's Pre-Collision Assist system. In cars, the system has a forward-facing camera and radar to detect vehicles, pedestrians and cyclists on the road and automatically applies the brakes if the driver doesn't respond to warnings. The new shopping cart has a sensor to scan for people and objects and applies the brakes when a potential collision is detected. Ford says the self-braking cart can help prevent damage to cars, store shelves and customers. They add it will help make grocery shopping less stressful for parents of little children.

Prototype & Intervention
Ford says one of the reasons they designed the cart was to demonstrate how important Pre-Collision Assist systems are to help drivers avoid accidents or reduce the impact. The new cart is a prototype at the moment. It's also part of a series of Ford "Interventions": applying automotive expertise to solve everyday problems.

34. New Wrist Tech to Stop Allergic Reactions: EpiWear

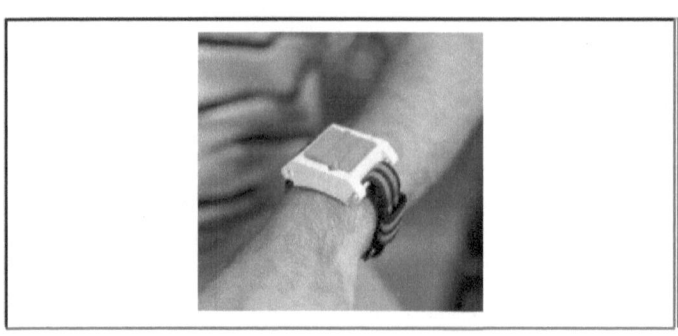

Source: Rice University Epiwear Watch to Halt Severe Allergic Reactions

Innovation from Rice University Engineers

The Rice University Engineering team calls this emergency medicine at hand at all times in case of a severe allergic reaction. It's an innovative, wearable epinephrine delivery device for your wrist that the developers say works much quicker than the traditional EpiPen, that's halts allergic reactions with its epinephrine shot.

Small, Foldable Watch

This is a small, foldable, watch-like device with a spring activated injector that will deliver a .3 millimeter dose of epinephrine into the body when needed. It's designed as a less obtrusive and more conventional alternative to the traditional EpiPen.

Function and Style

The Rice team is streamlining this prototype design to make it easier for people to keep what is life-saving medicine close at hand. This device was personally inspired by one of the researchers who has peanut allergies and significant experiences from that. The team is furthering their research to make this device less expensive, well styled and something that patients would really like to wear without broadcasting that they have a medical condition.

35. New Radar – Tiny, Cheap, Effective: Tracks Up to 12 Miles

Source: KAUST

Potential for Personal Security, Health, Drones, Self-Driving Cars

A tiny, low cost radar developed by an international team of scientists offers better navigation. Developers in Saudi Arabia and Finland say it could serve as an avoidance system for unmanned drones and self-driving cars. It could also be used as part of navigation systems for the visually impaired. They add it has potential for personal security and health care applications.

KAUST and VTT

The new radar device was created by researchers at the King Abdullah University of Science and Technology (KAUST) in Saudi Arabia and VTT Technical Research Center in Finland. The device weights less than 150g and is powered by a 5V battery. The team says it provides detailed information on the size, distance and speed of moving objects. The tracking range is up to 12 miles. Current radars are large and bulky. They tend to lose key detail information. This innovation doesn't. It's a low powered, small and portable radar that picks up on important small and big details.

36. Wristband Changes Color if Drink Spiked: Innovation from Young Entrepreneur

Source: Inventor Kim Eisenmann with her Xantus Drinkcheck Band

Protection Against Date Rape Drugs
It's called the Xantus Drinkcheck Band. A young German inventor, 25 year old Kim Eisenmann is the inventor. It's a wristband that can test drinks for date rape drugs including GHB - a drug that causes memory loss and drowsiness when mixed with alcohol.

Personal Innovation
Eisenmann invented the wristband after a friend was drugged and found injured after an event. To test a drink you stir it with a straw and drop a drip on the wristband. If the band turns blue, it's tainted with GHB. The wristbands are being sold in Germany. Eisenmann wants to start distributing them globally. A 2010-2012 US report estimated that 11 million women were raped under the influence of alcohol or drugs.

37. Pure Water, Green Bottle: Purifies Water by Ultraviolet Light

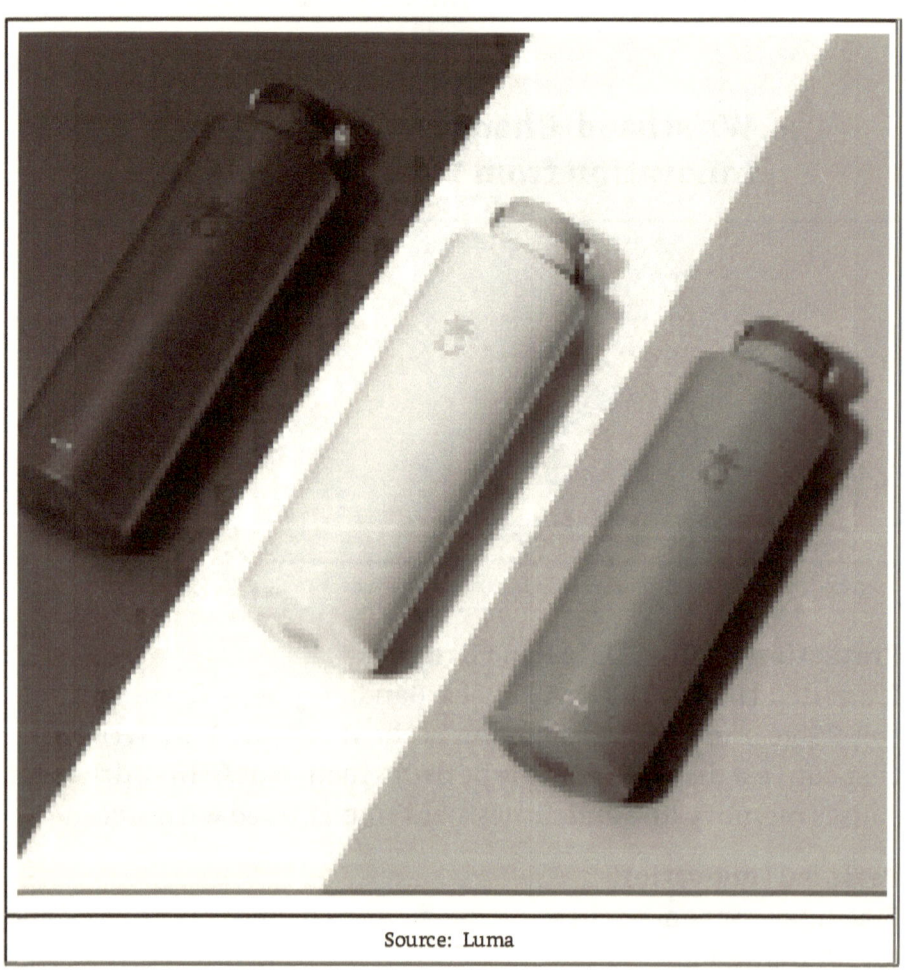

Source: Luma

Cuts Plastic Waste

The Luma bottle is a self-cleaning, reusable water bottle. It purifies your drinking water by UV light. All the while, cutting plastic bottle waste. The Luma technology for this green, innovative product is patent pending.

Destroys 99.9999% of Bacteria

The bottle contains a UV-C light at its base which is in constant contact with the water. In one minute it destroys 99.9999% of the bacteria in the water, according to Luma.

A Lot of New Innovation
The bottle contains a waterproof charging port and a Li-Polymer battery that can provide more than two months of cleaning cycles to purify the water on a single charge. This innovative product and Luma technology were developed by college students who started working on it in October 2017.

38. AI Enhanced Hearing: Evoke's Machine Learning Technology

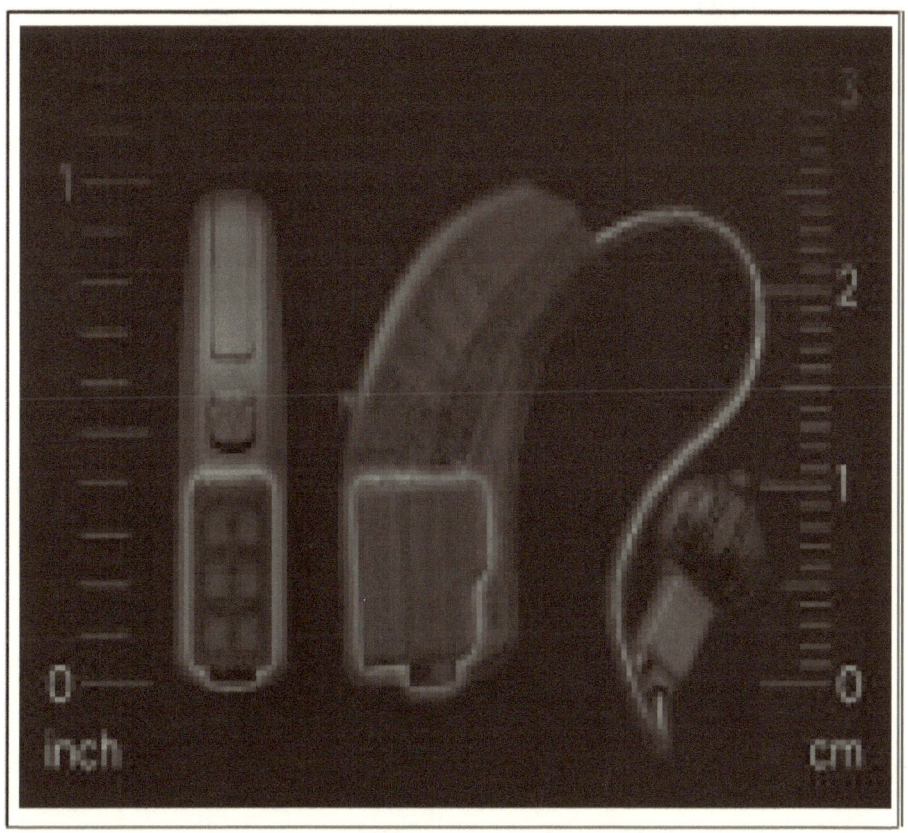

Source: Widex

Very Smart Hearing Aids

It's called Evoke. It's an AI enhanced hearing aid just developed that's state of the art. The smart device uses machine learning to optimize sound quality for the listener. With millions of people globally having impaired hearing, this new innovation has great humanitarian potential. This new technology was introduced at 2019 CES by Widex. The company calls it high definition hearing.

SoundSense AI Technology

The device's AI SoundSense technology pulls in real time data and uses it to adjust volume and customize settings for the individual. It's a battery-free hearing device.

Widex Energy Cell

Evoke is powered by Widex's Energy cell that combines oxygen and methanol and lasts for 24 hours. The recharge takes only 20 seconds and it's good to go for another day.

39. New Full Body 3D Motion Tracking Suit

Source: XSENS Motion Tracking Bodysuit

XSENS Innovation from the Netherlands

XSENS, based in The Netherlands, is a leading innovator in 3D motion tracking technologies. Their latest technology offers full body, 3D movement analysis by wireless data capture. They've introduced two new systems.

MVN Awinda

Their new MVN Awinda offers a full-body motion analysis that's optimized for sports and ergonomic uses. It deploys 17 wireless trackers wrapped around the performer such as in a headband, wrist, arm and leg bands. Movement data from them is transmitted up to 60 feet indoors and 150 feet outdoors to a backpack called an Awinda station where the data is streamed live or recorded.

Full Bodysuit

Another new product is the MVN Link bodysuit. It is a lycra bodysuit with wireless trackers built-in. It comes in 5 sizes and has a wireless range for data transmission of 150 feet indoors and 450 feet outdoors. The bodysuit transmits the data to a central data link that resends or streams the data. This is new technology to help athletes, assembly line workers and perhaps those undergoing physical rehabilitation to rebuild their limbs.

40. AI Beanie: Smart Hat to Helmet When Needed

Source: ANTI Ordinary AI Beanie

AI Shifts Hat from Soft to Helmet-Like Protection

The company ANTI Ordinary says this hat is as safe as a helmet and you'll find it your most comfortable helmet ever. It's called the AI Beanie and has soft, Merino wool knitted lining and a moisture repellent acrylic outer layer. What is between those two layers is the differentiator.

Non-Newtonian Fluids - Not Your Ordinary Hat

What is between the two layers is the company's proprietary blend of non-Newtonian fluids. The fluid particles remain soft and flexible in slow motion and become much harder when impacted. The inner layer of the hat becomes stiff enough to exceed safety standards while flexible enough to pack in a backpack and easily fit the contours of the head. It's deliverable in October 2019, priced at $125.00 and seems targeted particularly at outdoor winter sports.

41. Smart Curtin Removes Indoor Pollution: Air Purifying Textile

Source: Ikea's GUNRID Curtain & Mauricio Affonso

Innovation from Ikea Based on Plants

Ikea has developed a textile, first being deployed in curtains, that purifies the air from pollutants inside your house. Inside air pollution in homes is a huge problem. It causes 4.3 million deaths per year globally. In some parts of the world, indoor air pollutions is five times higher than outside air pollution.

GUNRID

The new material and curtains is called GUNRID. It's been developed by an international team of scientists, engineers, designers and specialists, including from Asian and European Universities The team is led by Ikea's Mauricio Affonso.

New Material

GUNRID works with light to filter toxins from the air inside the home such as odors and formaldehyde. The fabric has been treated with a photocatalyst that can break down common pollutants. It works in a way very similar to how plants filter out toxins. The chemicals in the curtain work with light or artificial light against pollutants. The company believes the coating can be applied to other material. GUNRID curtains will be available in

2020.

42. Scooter Sensation Stator: All Electric and Self Balancing

Source: Stator

A New Way to Scoot

It's called Stator and it's a self-balancing scooter, largely by its very wide tires. The vehicle is about to launch. The scooter is all-electric, green and is designed as an alternative means of urban travel. Stator is a California based company. The vehicle designer and company founder is Nathan Allen. The vehicle is being tested and readied for launch. No pricing on it yet. Here are a few specifics on it:

- Reaches maximum speeds of 25 mph

- 4 hour charge provides 20 miles of ride time
- 1.2 hour fast charge option
- Use a keyless RFID startup
- Breaks and throttle are on the right side
- The large rear tire holds the 1000 watt motor
- The front tire controls direction
- It weighs 90 pounds and can carry up to 250 pounds
- It folds for easier transport
- There are 3 programmable power settings: learner, intermediate, expert.   .

43. Double Folding Smartphone: This Phone Triples in Size

Source: Xiamoni, Double Folding Phone

Xiamoni's Double Folding Smartphone
This is a really cool new piece of smartphone innovation. The double folding smartphone from Xiamoni expands to triple its

size from a pocket phone to a tablet. This is a prototype but experts say there is nothing like it out there.

More to This - Shifting Screen Image

Xiamoni' s foldable phone shifts the image when opened to the size of the screen rotationally in an automatic fashion as you flip it. There aren't a lot of tech specs on this breakthrough phone as it is a prototype and under development. But it is something to be aware of as it goes to market. Thought you might enjoy knowing about this innovative development in smartphones.

44. KitchenAid Smart Display

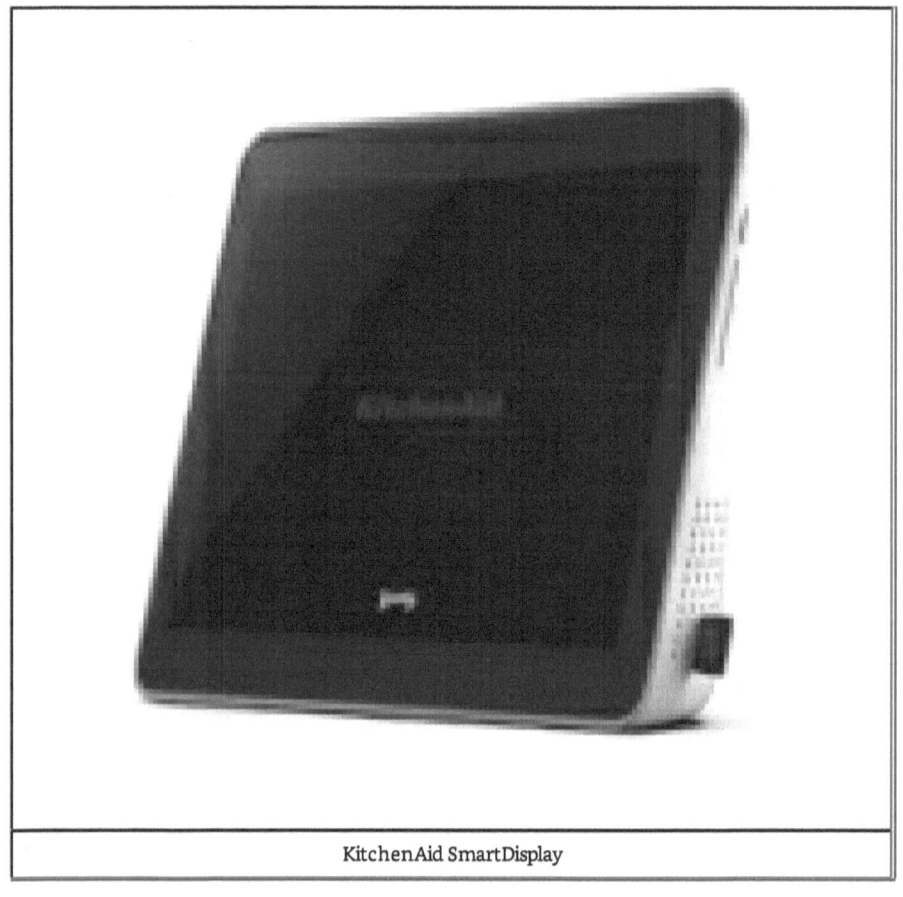

KitchenAid SmartDisplay

Getting a Lot of Interest

This smart display comes from a surprising source - KitchenAid. But they have created technology innovations in home appliances for many years. KitchenAid is owned by Whirlpool that owns another innovation company Yummly. Their new Smart Display takes all of that new tech to a new level. They've innovated a smart looking, voice activated touchscreen that has the Google Assistant onboard as well as Yummly app.

Smarthome Display

The ten inch display is water resistant and ready for kitchen duty. It's due out in the second half of 2019 and the pricing is expected to be from the $200 to $300 range.

Use by Voice Command Through Google Assistant

You can use the screen for many purposes like controlling your smart home gadgets, making video calls, doing search, watching videos, listening to music, all by your voice commands through Google Assistant.

MultiTasker & Your Recipe and Meal

This device is a multitasker, that can provide you video while providing you a recipe and cooking instructions at any time. It also offers an exclusive cooking app thru Wordpool's Yummly that gives you personal recipes and cooking instructions.

45. Smart Lighter to Quit Smoking: Slighter

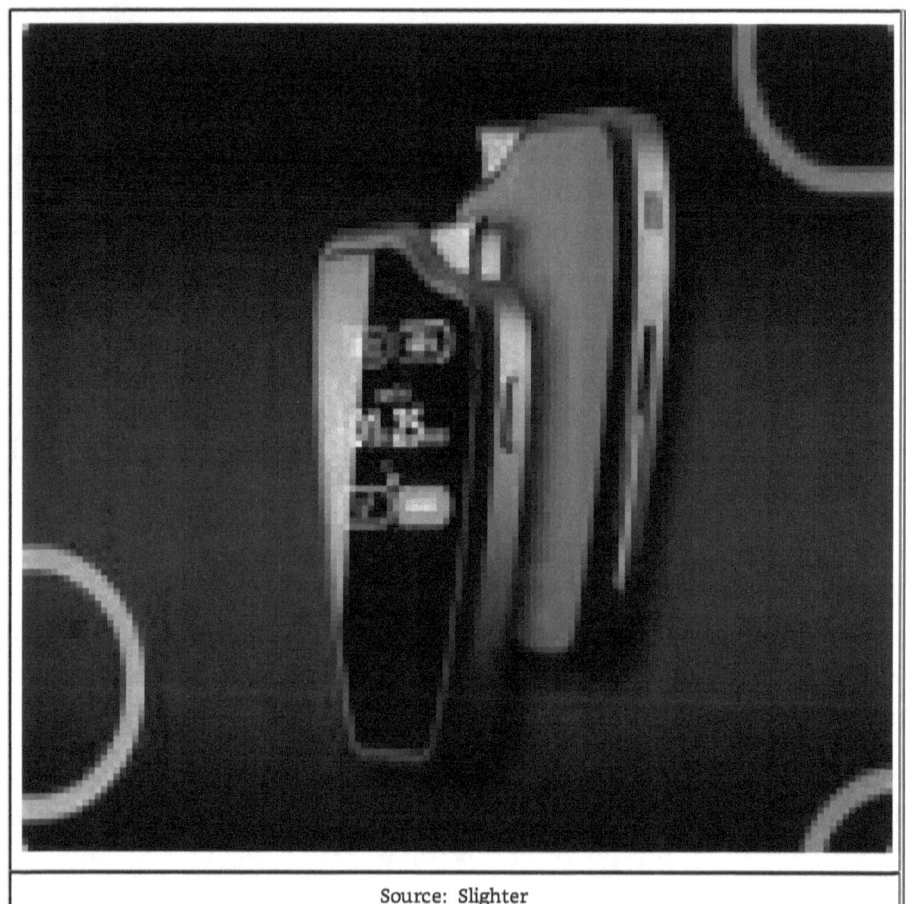

Source: Slighter

Innovation from Lebanon
Slighter is the latest, smart lighter that can help you quit smoking one cigarette at a time. The company is based in Lebanon. Their product offers a new way to quit smoking. It learns from your habits. It calculates how frequently you smoke and when you do so. It uses the data to customize a plan for each smoker.

Elegantly Simple Innovation that's Easy to Use
There's a lot of innovation and ingenuity in this smart lighter concept. It sometimes, on purpose, won't light up to help the smoker cut cigarettes eventually down to zero. The price is $129. It will start shipping in 2019. The French National Cancer Insti-

tute plans to do a trial run with it.

Communicates and Calculates

You can communicate and share the data on your progress to quit smoking on a phone app. The device also tells you how much money you've saved as you reach your quit smoking goals. Interesting and innovative technology innovation developed in Lebanon.

46. Very Smart Clock, Smart Home: Lenovo Smart Alarm Clock Control System

Source: Lenovo

Google Assistant, Time and a Lot More of a House Smart Control Center

This may be the world's smartest clock. Lenovo's SmartClock. All you need to do is say "OK, Google" to set alarms, get a weather update and see your schedule. You can control 5,000 of the most common smart home devices directly from it. You can even monitor other rooms in your home with it by using Nest smart cameras.

Interesting Innovation

There's a 4 display that's a touch screen. There's a light sensor that dims the screen while you sleep and gradually starts lighting it up 30 minutes before your alarm goes off. It also plays music. The price is $80.00 and Lenovo says it will be available in 2019.

47. Very Smart Mirror: Kohler Verdera -Voice Activated, Motion Detection, Alexa

Voice Lighted Mirror with Amazon Alexa

Kohler is known for its beautiful bath fixtures. At the Consumer Electronics Show 2019 in Las Vegas, Kohler introduced something very smart and unique. The Verdera Voice Lighted Mirror with voice control. The smart mirror has voice controlled light and motion detection that turns on the mirror lights when you walk into the room.

Alexa

The mirror is also equipped with Amazon's Alexa. That enables you to get messages, listen to the news or even shop as you get ready for your day. This smart mirror was a CES favorite innovation.

48. World's First AI Powered Pet Feeder

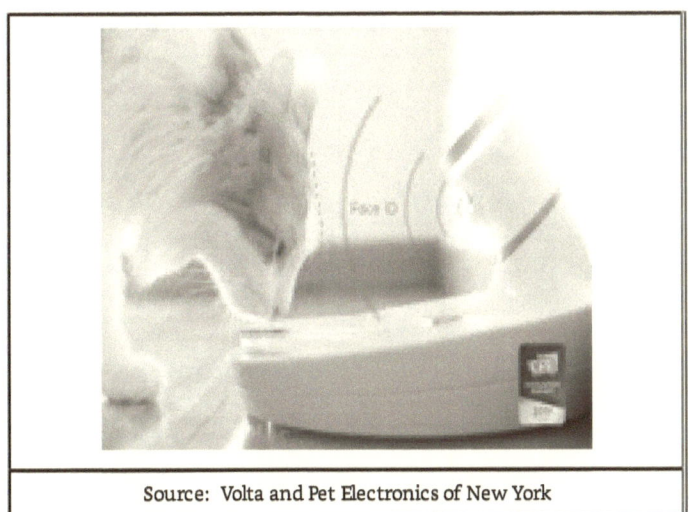

Source: Volta and Pet Electronics of New York

Innovation Machine that Caters to Your Pet

As a dog lover, I'm fascinated by this innovation for my German Shorthaired Pointer Rudi aka Snookie. He's just 1 years old, athletic and thin but a foodie. The new innovation is called Mookkie and it's been produced by the Italian Tech company Volta. It's a pet feeder with AI intelligence that recognizes each of your pets' faces and feeds them accordingly. My other pet is Biddie, a 6 year old Smooth Fox Terrier who is another foodie and a terrier about it. This AI pet bowl feeder just received the CES Innovation Award in the Smart Homes Category. Sounds interesting.

Right Portions
This device gives each pet the right amount of food. It works for dogs, cats and pets with special dietary conditions and needs. It's an innovation standout at CES 2019. It's going to be launched in September 2019 at a price of $189. It also sends you an alert on your cell phone that the bowl is empty and you need to refill. Interesting innovative tech for pet lovers. Volta developed this AI driven pet bowl with Pet Electronics of New York.

49. Communicative Cat Brush – Japanese Innovation

Source: Wataoka Industries

Neko-jasuri Cat Groomer

It's called a communications brush to help an owner and their pet cat better interact and communicate with each other. It's a metal brush - a pet hair brush inspired by another industry and company focused on metal files. This innovation comes from Japan, that is a nation of pet lovers. The company that innovated and is developing this product is Wataoka, with expertise in making metal files for 126 years.

It's in the Grooves

Here's how they've innovated their metal files into advanced pet grooming tools. The pet groomer has varying grooves patterned by the grooves in a cat's tongue. They say the grooves are great

69

for brushing and grooming your cat. And, the cats are so familiar with the sensation they love it because it simulates what they do for themselves.

Japanese Innovation At a Global Pet Scale
Nedo-jasuri means cat file. The Japanese government is profiling this innovation and showcasing it as a breakthrough.

50. Touching VR Objects - New Ultra Lite Gloves

Source: ETH Zurich

Enables Touching Virtual Objects for Real
It's the ultra-lite glove that's less than 8 grams per finger. The gloves enable you to feel, grasp and manipulate virtual objects.

The system developed by Swiss scientists provides extremely realistic feedback and could run on a tiny battery. That allows unparalleled freedom of movements for VR and AR. This is a big development in VR, AR tech.

Global Quest for This Tech
Globally for years, scientists, engineers, and software developers have been trying to achieve what the scientists at ETH Zurich and EPFL have done. It's the quest for tech that lets users grasp, touch and manipulate virtual objects while feeling they're touching something real.

Swiss Scientific Achievement
The gloves are called DextrES. They've been successfully tested on volunteers and will be presented at an upcoming international symposium. The gloves are composed of nylon with thin elastic strips of metal running over the fingers. The next step is to scale this up. And the biggest market is gaming, but also AR and healthcare, including training surgeons.

51. Heart Monitor Taped to Skin: Organic Sensor & Solar Cell

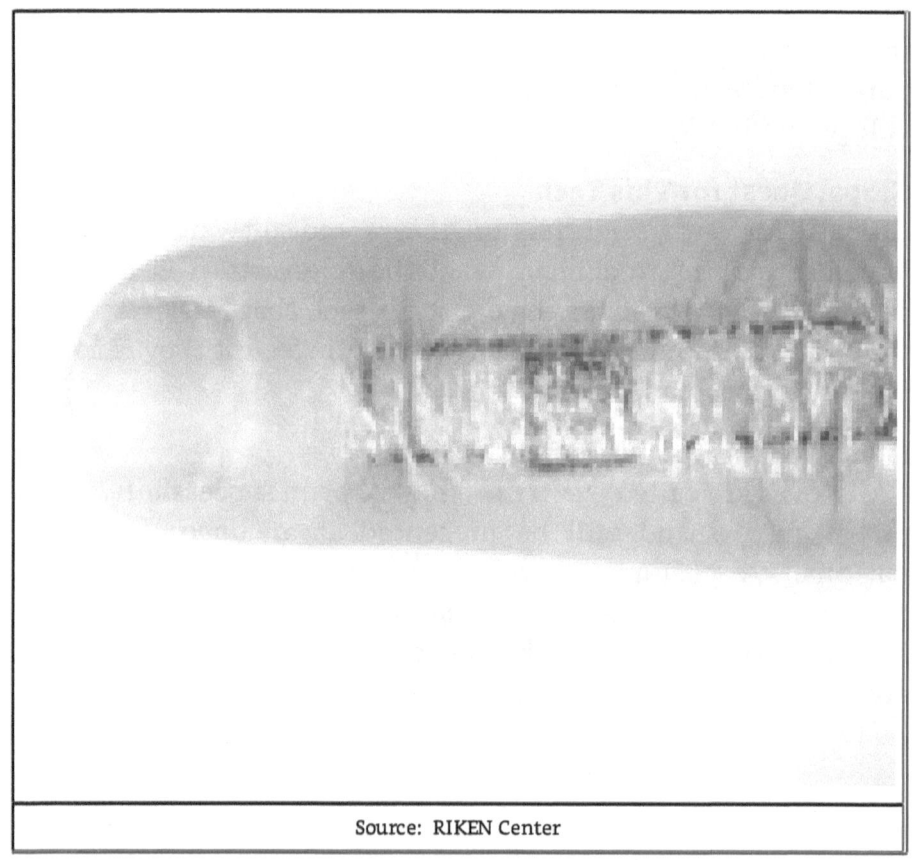

Source: RIKEN Center

Very Flexible, Taped Heart Monitor on Your Finger
Scientists have developed an organic sensor powered by sunlight
that performs as a self-powered heart monitor. The scientists
put a sensor device, an organic, electrochemical transitor, into a
flexible organic solar cell. The device successfully measures the
heartbeats of humans under bright light conditions.

New Generation of Self-Medical Monitoring
This monitor, developed by scientists at the RIKEN Center for
Emergent Matter Science and the University of Tokyo, is human
friendly, ultra flexible with a solar powered sensor. It can also be
used to monitor brain function. The development was reported
in the journal Nature. It's being hailed as the next generation of

self-medical monitoring. Such self-powered devices placed on the skin have great potential for medical applications.

52. LifeStraw - Personal Water Filter Technology

Source: LifeStraw

Don't Drink the Water

Dirty water kills millions of people every year. 2.1 billion people globally don't have safe water to drink. In fact, more people die from contaminated water than from anything else, including war and violence. A Danish entrepreneur has developed a high-tech straw that makes dirty water clean and drinkable.

Innovative Technology

The straw kills millions of microbes as the water passes through it. It works like a normal drinking straw. But it traps contaminants in a number of long, hollow fibers inside the plastic tube. The company says it traps 99.9% of parasites and 99.9999% of bacteria, including those that cause cholera and typhoid fever.

3rd World Use

This technology is of particular importance to developing countries where clean, drinkable water is a scarce commodity. By 2025, half of the world's population will live in regions where the demand for safe water exceeds supply.

Business Doing Good

LifeStraw says it's dedicated to doing good and giving back through retail. They say for every LifeStraw product purchased, a child in need receives safe water for an entire school year.

53. Rockabye Baby in Robo Cradle: SNOO Smart Sleeper

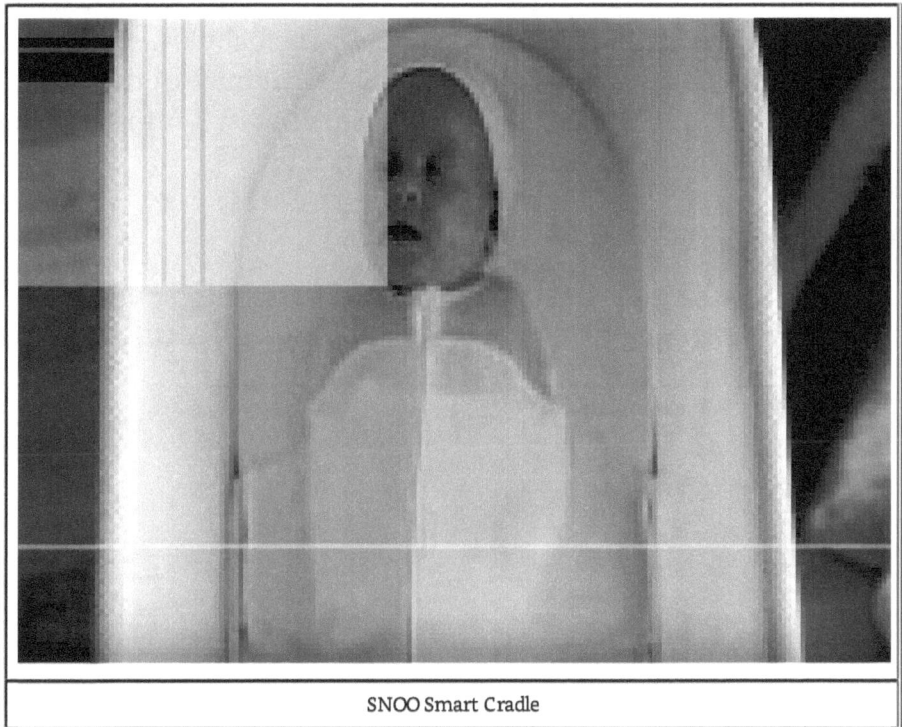

SNOO Smart Cradle

Sensors, Robotics and Microphones Concealed

It's called the SNOO and was designed by a Swiss innovator. Its sensors immediately pick up your baby's cry and automatically rocks the baby back to sleep. It customizes the best white noise and rocking motions for the child. The manufacturers claim it's able to add an hour or more to your baby's sleep every night. For parents of newborns that means an hour or more for them too.

Sleep Sack

Another important feature. It has a special sleep sack that attaches to the bed to swaddle the baby and importantly has the baby sleeping on their back, which is the safest position. Swiss designer Yves Behar calls his innovation "technology with a deep sense of humanity" as it performs an important function for parents and babies.

Minimalist Design

The minimalist design conceals the sensors, robotics and microphones. All the white, the SNOO encloses the baby in soft, washable material. It's designed as a personal helper but not to be used unsupervised. If the baby doesn't stop crying in 3 minutes, the SNOO turns off so a human attends to the baby.

54. High Tech Heated Clothing: Battery Powered Jackets To Stay Warm in Winter

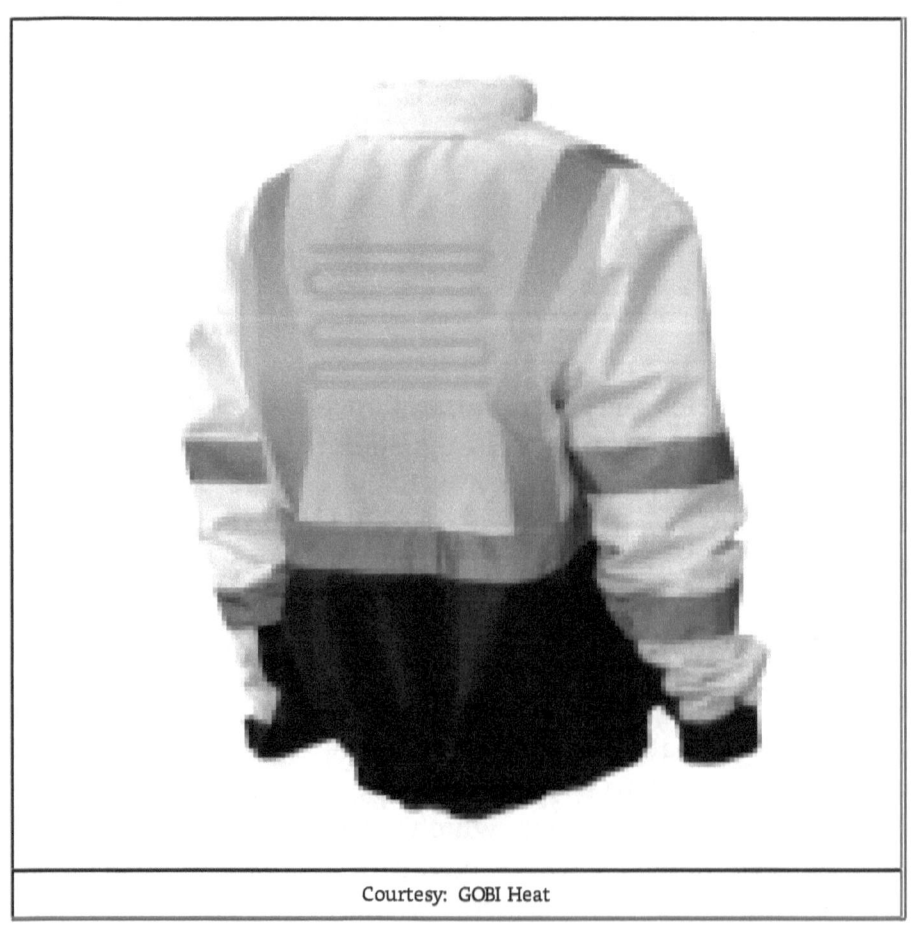

Courtesy: GOBI Heat

Wearable Innovation
Here's a case of practical, wearable innovation. A Utah company

GOBI Heat has innovated battery powered, heated clothing for outdoor winter activities. The apparel includes jackets, hoodies, vests, pants, gloves, socks. They come with a small battery that the company says will warm the clothing in 30 seconds against winter's chill.

7.4 Volt Small Battery

The 7.4 volt battery gives you up to 12 hour use on low heat and 7 hours on high heat. It has a one touch LED controller that can be set on low, medium or high depending on weather conditions. The company says their clothing is waterproof, surge-proof and have an emergency shut-off functionality. The jackets have 3 zone heating with 2, carbon fiber heating elements on the chest and a large one on the back. It gives new meaning to bringing the indoors, outdoors in freezing weather conditions.

55. 1st Rollable, Scrollable, Touchscreen Tablet: Inspired by Ancient Egyptian Scrolls

MagicScroll - a Rollable, Scrollable Screen

It's called MagicScroll. The device was created by computer engineers in Canada and it's a first. It's essentially a high resolution display. It can be rolled and unrolled around a 3D printed cylinder that has the device's computerized inner workings. The developers were inspired by and created it based on ancient Egyptian scrolls.

Canada Made for Many Uses like a Phone

MagicScroll was developed by researchers at Queens University Canada. The device's light weight and cylindrical body make it easy to hold in one hand. Importantly when rolled up, it fits into your pocket and can be used as a phone, dictation device or pointing device. Two rotary wheels at either end of the cylinder allow the users to scroll through information on the touch screen.

Goal: Any Device Can Be a Screen

Here's the ultimate point of the research. The researchers want to bring the MagicScroll device down onto something like a pen to demonstrate that virtually anything can be a screen with access to the internet.

56. Robot Vacuum Cleaners Start to CleanUp

Neato and Fido

Bots Do Windows Too

Robot vacuums & bots for other household chores like window cleaning are increasing in popularity. A number of factors are driving it. The technology is improving. There are growing concerns about allergens. And, for most of us, there is a lack of time to do the chores. There are a number of bot competitors including LG, Samsung, Neato, Evovacs, iRobot among others.

AIVI
Evovac has announced its vacuums are being equipped with AIVI, which is artificial intelligence and visual interpretation. This gives it object, environmental and spacial recognition so that it can do the job without human intervention.

Doing Windows
New cleaning bots were introduced at the IFA, which is Europe's biggest tech show. There's a new cordless, window washing robot. And a new dual action robot that's both mop and vacuum.

Neato and Fido
In the picture above, Neato seems to be Fido's best friend. Neato features zone cleaning and has advanced brushes to scoop up pet hair and allergens. It's also equipped with advanced LIDAR - LaserSmart mapping and navigation.

57. World's 1st VR Shoes & Gloves: Japan's Wearable VR Innovation

Taclim VR Gloves & Shoes - Japanese Innovation

They're called the Taclim Virtual Reality shoes and gloves. They're a global VR first. They have haptic feedback to enable users to feel the virtual world. As you can see in the photo, they work in tandem with the VR display.

Tactile Devices

Built-in tactile devices give the wearer instant haptic inter-action. The shoes and gloves generate the sensation of stepping onto the ground in a virtual space.

Come September

These products are relatively expensive - $1363 as posted in Japan Trends. They're the creation of Cerevo in collaboration with Nidec Seimitsu Corporation. This is fun and interesting VR tech I came across while researching Japanese innovation.

58. Paper Water Test: Quick, Cheap, Green Biosensor ID's Water Quality

Source: University of Bath UK Photo

Breakthrough Water Test

University of Bath UK scientists have developed a simple, paper-based test to determine if water is contaminated. It's potentially a simple, low cost way for developing countries to limit the spread of water-borne disease.

Simple as a Litmus Test

It's paper based like a litmus test to measure acid in water. But the new device is breakthrough. It consists of a microbial fuel cell (MFC). It's made by screen-printing biodegradable carbon electrodes onto a single piece of paper. MFC uses natural biology processes of electric bacteria- attached to the carbon electrodes - to generate an electric signal.

Don't Drink the Water

When the bacteria are exposed to polluted water, there's a change in the electric signals. That's the red alert not to drink. You're warned that the water is contaminated.

Lots of Potential

This is a revolutionary way to test water. It's point of use, cheap as $1, easy, rapid and green as the components are biodegradable. It could have a significant positive impact on developing countries. The scientists are working on linking the biosensor test to mobile devices like a cell phone through a wireless transmitter to make it very quick and user-friendly.

59. World's 1st Biodegradable Cooler: No Styrofoam, Environmentally Friendly

Source: Igloo Coolers

Biodegradable Igloo Cooler

As you head to the beach this summer, you might want to be aware of a new green gadget that Igloo has created. It's a cooler named ReCool and is the world's first cooler made of completely biodegradable materials. It can hold 16 quarts and retain ice for 12 hours. It's green, clean and cheap - just $10.00

Green ReCool

ReCool is made of organic molded paper pulp. To provide strength and water resistance, it's treated with ketene dimers. It can be reused multiple times and it's crumple resistant. This is a new piece of gadget innovation and designed as a way to improve the environment by getting away from Styrofoam.

www.ingramcontent.com/pod-product-compliance
Lightning Source LLC
Chambersburg PA
CBHW030726180526

45157CB00008BA/3060